CUTTING-EDGE TECHNOLOGY

ALL ABOUT STREAMING

by Rachel Kehoe

FOCUS READERS.
NAVIGATOR

WWW.FOCUSREADERS.COM

Copyright © 2023 by Focus Readers®, Lake Elmo, MN 55042. All rights reserved. No part of this book may be reproduced or utilized in any form or by any means without written permission from the publisher.

Focus Readers is distributed by North Star Editions:
sales@northstareditions.com | 888-417-0195

Produced for Focus Readers by Red Line Editorial.

Content Consultant: David Arditi, PhD, Associate Professor of Sociology, University of Texas at Arlington

Photographs ©: Shutterstock Images, cover, 1, 4–5, 7, 8–9, 13, 17, 20–21, 23, 24–25, 29; iStockphoto, 11, 14–15, 27; Red Line Editorial, 19

Library of Congress Cataloging-in-Publication Data
Names: Kehoe, Rachel, author.
Title: All about streaming / by Rachel Kehoe.
Description: Lake Elmo : Focus Readers, [2023] | Series: Cutting-edge
  technology | Includes index. | Audience: Grades 4-6
Identifiers: LCCN 2022035480 (print) | LCCN 2022035481 (ebook) | ISBN
  9781637394748 (hardcover) | ISBN 9781637395110 (paperback) | ISBN
  9781637395813 (ebook pdf) | ISBN 9781637395486 (hosted ebook)
Subjects: LCSH: Streaming technology (Telecommunications)--Juvenile
  literature.
Classification: LCC TK5105.386 .K44 2023 (print) | LCC TK5105.386 (ebook)
  | DDC 006.7/876--dc23/eng/20220826
LC record available at https://lccn.loc.gov/2022035480
LC ebook record available at https://lccn.loc.gov/2022035481

Printed in the United States of America
Mankato, MN
012023

ABOUT THE AUTHOR
Rachel Kehoe is a science writer and children's author. She has published several books and articles on science, technology, and climate change.

TABLE OF CONTENTS

CHAPTER 1
Many Types of Media 5

CHAPTER 2
How Streaming Started 9

CHAPTER 3
The Science of Streaming 15

HOW IT WORKS
Streaming Services 18

CHAPTER 4
Streaming Today 21

CHAPTER 5
Pros and Cons 25

Focus on Streaming • 30
Glossary • 31
To Learn More • 32
Index • 32

CHAPTER 1

MANY TYPES OF MEDIA

You and a friend are talking after school. She tells you about her new favorite song. But she can't remember all the words. So, she pulls out her smartphone and opens an app. Your friend types the name of the song. Instantly, the music starts to play.

Streaming makes it easy for people to listen to and share music.

Your friend sends you a link. Now you can listen, too.

That evening, it's your turn to do the dishes after dinner. You listen to an audiobook as you work. Once again, you use a smartphone. An app takes you to an online library. You choose the book you were listening to yesterday. The app begins right where you left off.

Before bed, you decide to watch a show. You use your dad's laptop to log in to a website. It suggests several new shows you might like.

All these activities are examples of streaming. Streaming is a way to play different types of **media** by sending

Smartphones, tablets, and computers are some of the many devices that can stream media.

information over the internet. People stream TV shows, movies, video games, music, and more.

Streaming makes viewing and listening easy. Streaming services let users pick what content they see or hear. Users can access this content at any time of day. And they can use many different devices.

CHAPTER 2

HOW STREAMING STARTED

Music, movies, and TV have been around for years. But streaming was only possible after the internet was invented. The internet is a huge network. It connects millions of computers and devices. It allows them all to share information, or data.

The internet lets devices all over the world send and receive information.

In the 1990s, people began using the internet to **broadcast** music and shows. In 1993, a band streamed a video. Viewers could see and hear the band play music. In 1995, a company streamed a baseball game. To listen, people could

THE EARLY INTERNET

The internet was created in 1969. It linked four computers together. They could send information to one another. Gradually, more computers were added. At first, only colleges and government workers could use the internet. But by the 1990s, that had started to change. More people owned computers. And the first websites were being created. Today, the internet includes devices all over the world.

Wi-Fi makes streaming easier. It uses radio waves to send data through the air instead of along wires.

download a new type of **software**. It let them hear live radio describing the game.

By 1997, more companies were looking to stream media. Many focused on videos. Videos combine images and sound. So, they use much more data than just audio does. To stream videos, computers had to send and receive lots of information very quickly.

11

At the time, most people had dial-up connections. This method used telephone lines to access the internet. It was easy to set up. But it had limited bandwidth. That meant it couldn't send much data. As a result, streaming was difficult and slow. In the early 2000s, broadband connections began replacing dial-up. This type of connection was much faster. It made streaming easier.

In the early 2000s, people began using websites to handle streaming. Users no longer needed to download programs. Instead, they could play media stored on **servers**. Computers could receive small bits of information at a time. Part of a

Many people use streaming to create and share videos of themselves.

song or video could start playing while the rest was being sent.

YouTube launched in 2005. This website let users watch and **upload** videos. It quickly became popular. Other streaming websites soon followed. Netflix introduced streaming services in 2007. It let people watch movies and TV shows. Other websites were designed to play music or podcasts.

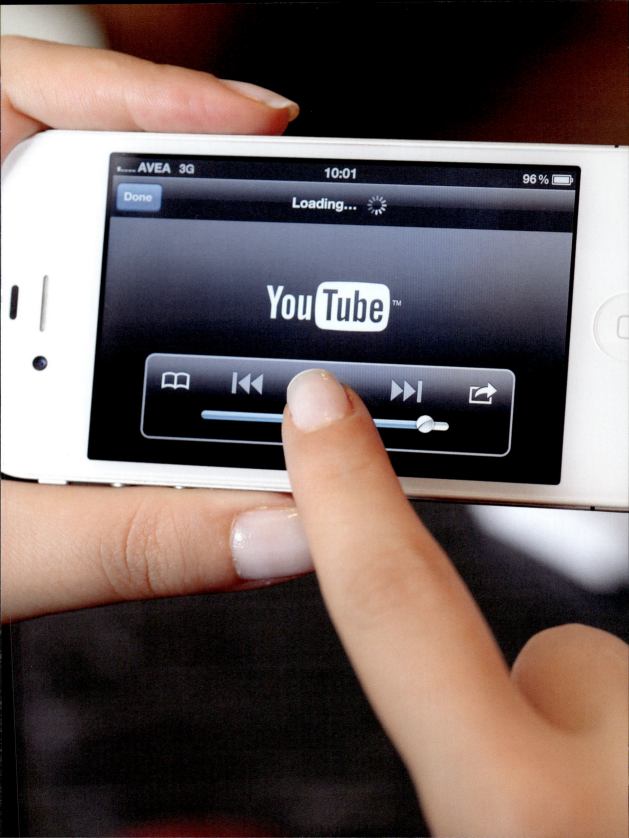

CHAPTER 3

THE SCIENCE OF STREAMING

Streaming works by splitting sounds or videos into small bits called packets. These small bits of data are sent from one computer or device to another. A computer puts the packets together to re-create sounds or images.

To download a file, a person must wait for all of the packets to arrive.

Users may see a loading symbol while devices wait for packets to arrive.

The file is then saved on their device. But streaming sends just a few packets at a time. Packets are played soon after they arrive. Then they're deleted. Nothing is saved to the person's device. This makes streaming one of the quickest ways to receive content.

LIVESTREAMING

Livestreaming lets people watch events as they happen. However, most livestreams have a slight delay. It takes time for the first few groups of packets to be sent over the internet. Livestreaming is often used for gaming and sports events. People can livestream concerts, too. During a livestream, viewers can interact. Websites often have places where people can post messages.

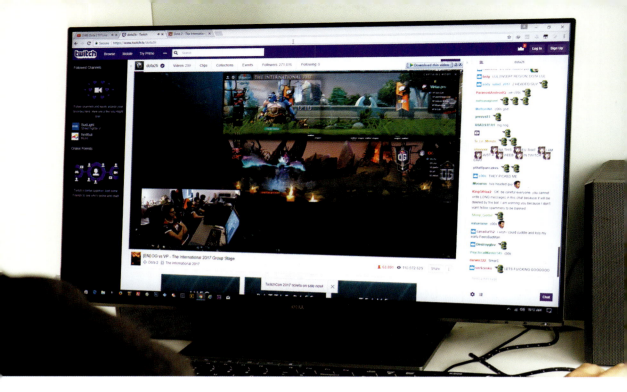

Twitch is one website that lets people watch and comment on livestreams of video games.

Streaming is similar to the way TV and radio stations use radio waves to send signals. However, streaming gives listeners more control. Users can often choose what and when they watch. They can also pause or go back to replay certain parts.

HOW IT WORKS

STREAMING SERVICES

By the early 2020s, many companies offered streaming services. Netflix remained one of the most popular. Other top choices included Disney+, Hulu, and Amazon Prime Video.

Some streaming services mainly rely on websites. YouTube TV is one example. Other services, such as Spotify, use apps. These programs are saved to users' devices. But the content they play is not saved. For example, each Spotify user downloads an app to a computer or smartphone. Then, the user can play millions of podcasts and songs.

Payment options also vary. Some services let people sign up for free. But these users must often watch or listen to ads. And they may not be

TOP STREAMING SERVICES

This chart shows which streaming services were most popular among people who paid for subscriptions. Figures are from July 2020 to June 2021.

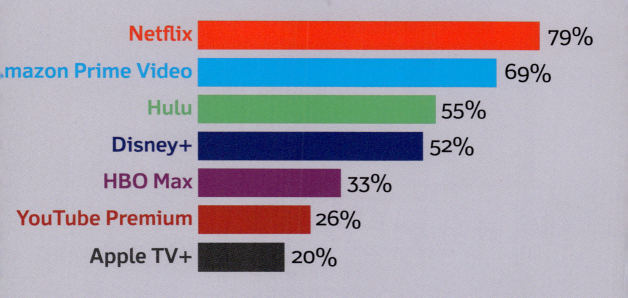

able to access all the service's content or features. For no ads or more content, people usually have to pay. Many services charge a monthly fee.

Some services focus on a certain type of content. Crunchyroll is a popular way to watch anime. And Acorn TV has many British shows.

CHAPTER 4

STREAMING TODAY

Streaming has changed how people watch and listen to media. Instead of buying songs, people can listen online. By 2022, streaming was more popular than traditional TV in many US households. As a result, many TV stations now have ways to watch their shows online.

By 2022, 80 percent of US households used at least one streaming service.

Streaming has also changed how people create content. People can watch shows and hear music from around the world. Users make some of this content. Other content is made by streaming services. Companies often track what

STREAMING WORLDWIDE

Not all people have access to streaming. Some countries have limited bandwidth. People in rural areas also tend to have slower connections. As a result, streaming doesn't work as well. Some companies and governments want to solve this problem. They work to provide high-speed internet to more places. Other places try to limit people's access to streaming. For example, some governments ban certain apps or websites.

China's government has banned people from watching YouTube since 2009.

people watch and like. Then they make similar movies and shows. By the early 2020s, their work had won several major awards.

Streaming can even make media interactive. For example, some shows let viewers click to change their endings. Shows could become even more creative in the future.

🔒 twitch.tv   ↻

All Games - Twitch

Mobile   Try Prime   ···   🔍 Search

Games   *Communities* `Beta`   Popular   Creative

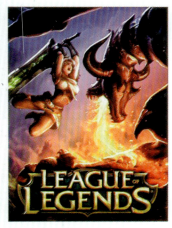

League of Legends
86,569 viewers

PLAYERUNKNOWN'S BA...
77,216 viewers

Hearthstone
59,358 viewers

Dota 2
29,872 vie...

Overwatch
18,757 viewers

Co...
18,127 viewers

TwitchCon 2017 tickets on sale now!   ✕

This site uses cookies. By continuing
to browse the site, you are agreeing   ✕
to our use of cookies. Review our
Cookie Policy for more details.

IRL
14,413 vie...

CHAPTER 5

PROS AND CONS

Streaming is convenient and easy. But it can also create problems. For instance, most streaming companies collect data about their users. They track what each user watches. This data helps companies recommend other shows the user might like. Companies also use the data to decide what new shows to make.

Streaming websites often use cookies to collect and track information about users.

25

However, many people worry about users' privacy. Some users don't want their data to be collected. They are concerned that companies could sell or misuse it. But they can't opt out.

In addition, streaming makes it easy to binge-watch. Users can watch many episodes in a row. Young people are especially likely to binge-watch. Spending too much time streaming can take time away from other things, such as schoolwork and friendships.

Pirated content is another problem. People may use illegal websites to play music or watch shows for free. These websites make it easy for people to watch.

Studies have found that too much screen time can make people feel lonely or anxious.

But the people who made the content don't get paid. Plus, illegal streaming can be dangerous. The websites may not be secure. **Hackers** could steal data or access people's computers.

Livestreaming is a great way for people to connect and share moments. It can

27

also improve **accessibility**. For example, travel is hard for some people. Streaming makes it easier for them to participate in events. Plus, many livestreams include captions or transcripts. People who are hard of hearing can read this text and understand what's happening.

STAYING CONNECTED

In 2020, the COVID-19 **pandemic** forced schools to close. Sporting events got canceled. Theaters and cinemas shut down. To avoid spreading germs, many people stayed home. But streaming helped them stay connected. People watched livestreams of events they couldn't go to in person. They also streamed movies and shows. With streaming, creators could still reach audiences.

Streaming can help students attend online events and classes.

However, livestreaming does have risks. Many livestreams are recorded. They can be saved or shared. Viewers should check the safety settings before they comment. And they shouldn't say anything they don't want to be public.

FOCUS ON
STREAMING

Write your answers on a separate piece of paper.

1. Write a paragraph describing the main ideas of Chapter 3.

2. Which potential problem with streaming do you think is most serious? Why?

3. Which type of internet connection made streaming faster and easier?
   - **A.** dial-up connections
   - **B.** phone line connections
   - **C.** broadband connections

4. Why should livestream viewers be careful when posting comments?
   - **A.** Their comments cannot be seen by other viewers.
   - **B.** Their comments may be spread or saved after the livestream ends.
   - **C.** Their comments cannot be spread or saved after the livestream ends.

Answer key on page 32.

GLOSSARY

accessibility
Making things easy for more people to use or understand.

broadcast
To send out radio or TV signals.

download
To copy data from one computer system to another.

hackers
People who illegally gain access to information on computer systems.

media
Different ways of sharing information. Books, music, and videos are all types of media.

pandemic
A disease that spreads quickly around the world.

servers
Computers that make data or programs available to other devices.

software
The programs that run on a computer and perform certain functions.

upload
To move or copy a file from a device to a website.

TO LEARN MORE

BOOKS

Green, Sara. *Netflix*. Minneapolis: Bellwether Media, 2018.
Harris, Duchess. *Your Personalized Internet*. Minneapolis: Abdo Publishing, 2018.
Swanson, Jennifer. *How Does the Internet Work?* Mankato, MN: The Child's World, 2022.

NOTE TO EDUCATORS

Visit **www.focusreaders.com** to find lesson plans, activities, links, and other resources related to this title.

Answer Key: 1. Answers will vary; 2. Answers will vary; 3. C; 4. B

INDEX

apps, 5–6, 18, 22
audiobook, 6

bandwidth, 12, 22
binge-watch, 26
broadband, 12

computers, 9–12, 15, 18, 27

download, 11–12, 15, 18

internet, 7, 9–10, 12, 16, 22

livestreaming, 16, 27–29

movies, 7, 9, 13, 23, 28
music, 5, 7, 10, 13, 22, 26

packets, 15–16
privacy, 26

servers, 12
smartphone, 5–6, 18
software, 11
streaming services, 7, 13, 18–19, 22

TV, 7, 9, 13, 17, 21

videos, 10–11, 13, 15